Jürgen Dollmann

Semiotic analysis of building structures of the Wimpfener Königspfalz

AF301551

Jürgen Dollmann

Semiotic analysis of building structures of the Wimpfener Königspfalz

A contextualization of history, medieval theology, philosophy and architecture

ScienciaScripts

Imprint

Any brand names and product names mentioned in this book are subject to trademark, brand or patent protection and are trademarks or registered trademarks of their respective holders. The use of brand names, product names, common names, trade names, product descriptions etc. even without a particular marking in this work is in no way to be construed to mean that such names may be regarded as unrestricted in respect of trademark and brand protection legislation and could thus be used by anyone.

Cover image: Provided by the author

This book is a translation from the original published under ISBN 978-620-0-44882-8.

Publisher:
Sciencia Scripts
is a trademark of
Dodo Books Indian Ocean Ltd. and OmniScriptum S.R.L publishing group

120 High Road, East Finchley, London, N2 9ED, United Kingdom
Str. Armeneasca 28/1, office 1, Chisinau MD-2012, Republic of Moldova, Europe
Printed at: see last page
ISBN: 978-620-7-90064-0

Contents

1. Introduction and thesis

Many people in the High Middle Ages tended towards a symbolic-allegorical interpretation of their sensually perceptible environment, as can be explained in this book from various sources and perspectives. This world view was strongly related to Christianity, which characterised the Middle Ages: behind the surface of the visible world, the work of God was assumed, which could not be directly revealed to people. However, the possibilities of recognising truths beyond the empirical world were already analysed by theologians and philosophers of scholasticism in late antiquity, but above all in the High Middle Ages, whereby reference was also made to the writings of ancient philosophers, above all Plato and Aristotle, in addition to the biblical texts. From the combination of these sources, the conclusion was drawn, among other things, that God offers people *signs* to gain supra-empirical knowledge. This takes place in symbols or allegories,[1] i.e. in words or images as well as in events that need to be interpreted.

The assumption of a deeper truth behind the surface of the sensual world had visible effects on medieval works of art, which therefore also had symbolic-allegorical elements and should certainly not only be viewed from an aesthetic perspective[2] . Both paintings and the architecture of churches or royal and imperial palaces can therefore also be analysed from this perspective. In art history, this is a scientifically recognised method in the context of iconography,[3] medieval architecture has only been analysed

---

[1] For the terminology of symbol and allegory, see section 3.2

[2] In contrast to the use of the term "aesthetics" by Georg Wilhelm Friedrich Hegel, for example , it does not refer to the idealistic reduction of beauty and art, but is understood in the sense of the Greek term *aisthesis* as perception and sensation (Cancic and Mohr 1988, 131).

[3] For examples, please refer to Poeschel 2007, passim.

from this perspective for a few years.[4] As the predominant political conflict in the High Middle Ages was the dualism of the empire and the papacy, and thus the question of whether the primacy belonged to secular or spiritual power, the symbolic-allegorical analysis of medieval buildings will also have to be analysed in terms of this conflict. With regard to the royal palace in Wimpfen[5], there has been no work on these aspects to date, and this gap is to be filled with the present analysis.

In addition to historical contextualisation and the theology and philosophy of scholasticism, the methodological focus is on general aspects of art and architectural history, as well as the semiotics of Ferdinand de Saussure. Umberto Eco's analyses of medieval art and beauty provided essential and helpful suggestions. The core thesis of the present work can be derived from the sources and the methodology: The conspicuous numerical rhythm of the sequence of arcades, the equally conspicuous order of columns and the ostentation of the Wimpfener Königspfalz are regarded as carriers of meaning. These structures refer non-verbally to contextual, socio-cultural events and ideas in the building period, which was characterised by Christianity and which provide insights beyond the mere historical facts.

Sections 2 and 3 of this thesis set out the theological, philosophical and aesthetic context in which the Wimpfener Pfalz can be categorised on the basis of its construction period. Section 4 describes the impact of this context on medieval architecture in general. Section 5 provides an introduction to semiotics, before analysing the Wimpfener Pfalz in terms of these aspects in section 6. Finally, the justified criticisms of this thesis are reflected upon. However, reasons are given as to why the necessity of

---

[4] An overview can be found, for example, in Binding 1998 and Bandmann 1998, both authors are referred to several times in the present work.
[5] The town was only given the name *Bad* Wimpfen in 1930; the historical name *Wimpfen is used in the text.

such an analysis is nevertheless historically and socio-culturally meaningful and plausible.

2. The philosophical and theological context in the Middle Ages

Medieval theology cannot be understood without basic concepts of Greek philosophy. Only a very superficial overview can be presented here; for a more in-depth understanding, please refer to the bibliography.

In the 4th century BCE[6] , *Plato* contrasted the transient world of appearances with the unchanging *ideas*. The sensually perceptible world is a place of often illusory appearances that conceal the unchanging reality behind them (Papineau 2006, 77 f.). According to Plato, the ideas have an independent existence. However, since the world of appearances is an image of the world of ideas, the world of ideas can be recognised from the formation of analogies. (Aster 1968, 6468). A few years later, *Aristotle* developed his philosophy, which was more worldly in that it did not deal with the idea, but with the concept of substance: matter (hyle) must be accompanied by form (morphe), which ultimately creates the individually real; matter is thus *informed*. For example, the material *clay* becomes a *brick* through the form, the material *brick* is formed into a *wall*, this form in turn becomes a *house*. This *substance-form principle* (only later referred to as *hylemorphism*) can then be logically traced back to an abstract first, pure form, a "first mover" that is itself motionless (Aster 1968, 87). Another important element of the Aristotelian

Philosophy is the distinction between the *existence of* things on the one hand, i.e. their *being*, and their *essence* on the other. The philosophies of Plato and Aristotle were taken up and transformed theologically in

---

[6] In this work, the years *B.C. - before Christ -* are given as *B.C. - before our era - in a* religiously neutral manner. Years without further specification correspond to *n. Chr. - u. Z. - our era.*

Christianity:

A decisive figure in patristics was Augustine (354-430), who referred to Plato's philosophical thought system (Aster 1968, 129). His reception of Plato, a form of Neoplatonism, was the first to "lend language to the philosophical thought of the Christian world age" (Gadamer 1965, 272). For Augustine, the absolute truth lying behind the world of appearance is "the being of God" (Aster 1968, 129 f.) Analogously, a soul in the Christian sense can be seen behind individual human life in the Neoplatonic sense.

It was only in the early Middle Ages that Aristotle's works were rediscovered through contact with Islam in the West[7] , for example during the Moorish immigration to the Iberian Peninsula and later during the Crusades. Muslim philosophers such as Avicenna (9801037) and Averroes (1126-1198) played a decisive role in this process, passing on Aristotle's texts to the West (Franzen 2006, 212 f.). The Christian philosophers and theologians of the West were impressed by the clarity of this thought and endeavoured to combine Aristotelian philosophy with Christian doctrine. Anselm of Canterbury (1033-1109) is regarded as the founder of this school of thought, known as *scholasticism* (Franzen 2006, 211). Scholasticism provided the Christian Church with its official philosophy through Thomas Aquinas (1225-1274), which is still canonically valid in the Catholic Church to this day (Gadamer 1965, 272). According to Thomas Aquinas, the "first form" developed by Aristotle is *God*, the "first mover" (Aster 1968, 160). The Aristotelian distinction between *existence* and *essence* also had a decisive influence on Christian thought in the High Middle Ages. This can be illustrated, among other things, by the fact that at

---

[7] Here, the "West" is to be referred to non-normatively and historically, initially to the European continent - since the 15th century also to North America - and in the Middle Ages it was contrasted with the *East*, the *Orient*. The problems of Orientalism and post-colonial theories cannot be discussed further here.

the 4th Lateran Council in 1215, the so-called *transubstantiation*, i.e. the transformation of the host in the Eucharist, was decided under scholasticism: on the basis of the priest's authority, the host is not transformed from *existence*, but from *essence,* from *being* to the body of Christ. (Franzen 2006, 211). Finally, in his work *Summa theologiae,* Thomas Aquinas developed an overall presentation of Catholicism on "[...] the philosophical theological basis of Christian Aristotelianism [...]" (Franzen 2006, 213). The integration of Greek philosophical thought also led to a reinterpretation of numerous New Testament texts:

Umberto Eco refers to a passage from St Paul's letters that is characteristic of the Middle Ages (Eco 1993, 79 f.): "For now we see by means of a mirror indistinctly, but then face to face." (1 Cor 13:12).[8] Correlating with the - albeit fundamentally different - approaches of Plato's *theory of ideas* and Aristotelian *hylemorphism*, this quotation from the First Epistle to the Corinthians was able to philosophically plausibilise Christian thought: For people of the Middle Ages, the world was full of meaningful references to God behind superficial, sensory perception, resulting in double meanings. These

[8] All biblical quotations (except below a quotation from the *Wisdom of Solomon* as an apocryphal text) were taken from the Elberfelder Studienbibel mit Sprachschlüssel (2005), see Literature.

Eco describes this way of thinking as a "[...] symbolic-allegorical view of the world." (Eco 1993, 79). The aim of believers in the Middle Ages was to fathom the divine truth behind things from an emic point of view by drawing conclusions by means of analogies. "But the thing is not what it seems to be; it is a sign of something else" (Eco 1993, 81). This way of looking at the world, which was full of meanings, clues and double meanings, had an impact on aesthetic perception with regard to the aspect of *beauty.*

3. On the aesthetics of the Middle Ages

3.1 Beauty in the Middle Ages

As a result of the background explained above, *beauty* in the Middle Ages did not primarily refer to beauty that could be grasped by the senses; the concept was extended to "inner beauty" (Eco 1993, 23). According to Eco, the Middle Ages mistrusted external, sensually perceptible beauty. However, there are analogies between beauty that can be perceived by the senses and beauty that cannot be perceived by the senses, which, as explained above, should be recognised by believers. A further philosophical aspect is significant in this regard: scholasticism defined the general properties of everything that exists, the so-called *transcendentals,*[9] which included the good and the true: the world created by God is *true* and must also be *good* (Eco 1993, 37 f.). There were now extensive discussions about whether *beauty* should also be counted among the transcendentals. In the *Summa fratris Alexandri*[10] , the beautiful is equated with the true: True, good and beautiful are convertible and only differ for the mind. Truth is the nature of the form in relation to the interior of the thing, beauty is the nature of the form in relation to the exterior (Eco 1993, 42-48). This is also interpreted by Albertus Magnus: The good is inherent in the beautiful because the beautiful has the same substrate as the good. So if the eye recognises a thing as *beautiful*, the intellect must try to recognise the *good* and *truth* behind it. This is also very clear from a quote of the time from Hugh of St Victor, a 12th century mystic:

---

[9] The complexity of the transcendental doctrine of scholasticism cannot and need not be explained in this text. Reference may be made to Diemer (1970) page 235 and Eco (1993) pages 34-48.

[10] The text was attributed to Alexander of Hales (1185-1245), but it is the work of three French authors in which the problem of the transcendental nature of beauty was clarified in its particularity (Eco 1993, 42).

All visible objects are placed before our eyes to designate and explain invisible things, and they teach us through the eye in a symbolic, that is, in a figurative way...Because the beauty of visible things consists in their form...the beauty of visible things is an image of the beauty of invisible things. (Quoted from Eco 1993, 91-92, punctuation by Eco).[11]

From today's perspective, we have certain difficulties in empathising with this symbolic thinking, so here are some general aspects of the concept of the *symbol*.

3.2 <u>General aspects of medieval symbolism</u>

The terms *symbol* and *allegory* were used synonymously in the Middle Ages; their differentiation only began in the 18th century, probably going back to Goethe's aphorisms (Eco 1993, 85).[12] For this reason, Eco formulates the *symbolic-allegorical universe* of medieval man. According to the theologian and art historian Gerd Heinz-Mohr, a certain dimensional reduction becomes visible in the "insecurity and unease of today's Western-civilised man towards the symbol" (Heinz-Mohr 1971, 7):

He [note: man] possesses a wealth of special knowledge, which he holds in his hand as fragments and realises independently of one another for a given reason, but hardly has the "unifying bond" that joins them together and orders them meaningfully [...] (Heinz-Mohr 1971, 7, inverted commas by the author).

Heinz-Mohr refers to the etymology of the term *symbol*: it can be traced back to the Greek συμβάλλειν *(symbállein)*, which can be translated as *bringing together,* "[...] to a meaningful place [...]" (Heinz-Mohr 1971, 9) can be translated.[13] In this use, the symbol must be understood as

---

[11] The original source in Latin was not used here, but Umberto Eco's translation.

[12] According to these aphorisms, the allegory translates the *concept of a **phenomenon*** into an image; the symbol is the *concept of an **idea that is*** shown in the image (Eco 1993, 85).

[13] Critically, it must be noted that etymological aspects cannot be used to attribute a recent

something non-static, as it requires active participation in order to be grasped. This coincides with Eco's interpretation of the term symbol, which assumes an "incongruity of the symbol in relation to the symbolised thing" (Eco 1993, 83 f.). There is thus no identity between *symbol* and *thing,* but rather a proportional relationship that leads to aesthetic pleasure in the context of an interpretative endeavour. Eco points to an argument that is particularly important in the approach of the present work: today's people may not share this way of thinking,[14] but in our interpretation of medieval artefacts we must take into account that at that time - the *symbolic universe* - everything had its own place, everything corresponded to each other, and a corresponding harmonious combination of visible and invisible truths was sought. The philosophical and theological background was discussed above.

3.3 The symbolism of numbers in the Middle Ages

The significance of numbers goes back to Pythagoras and his school: the Pythagoreans were the first to investigate the mathematical relationships that organise musical tones, with physical proportional relationships playing a central role (Eco 1993, 51 f.). Biblically, a quotation from the Apocrypha attributed to Solomon becomes decisive for medieval aesthetics (Eco 1993, 34), which says of God's creation: "But you have ordered everything according to measure, number and weight" (Wisdom 11:21).[15] This series of concepts of measure, number and weight in the wisdom teachings of Israel, developed between 80 and 30 BCE, can also be found in Greek writings by Plato and later by Philo of Alexandria (Binding 1998,

---

meaning to a term. A post-structuralist approach would prohibit this.

[14] However, this argument must also be related to the creation of the Italian version of Eco's cited work in 1987. In the last 40 years or so, there has been a worldwide increase in religious studies towards an individualised, holistic spirituality (Dollmann 2021, 51-65).

[15] This quotation, which is not included in the Elberfelder Study Bible as an apocryphal scripture, was taken from a Luther Bible. See literature: Die Bibel: Nach der Übersetzung Martin Luthers.

419). St Augustine also comments on the *Wisdom of Solomon* that measure and number reflect a divine order (Binding 1998, 190; Eco 1993, 34). For Augustine, measure, number and order are to be found in the most perfect way in God himself (Binding 1998, 191). This can be illustrated by the fact that the Creator God is often depicted with a compass, as can be seen in the Austrian National Library, for example: *God, Architect of the Universe.*[16] According to the late antique philosopher and theologian Boethius (ca. 480-525), *number is* the basic principle of all things, the laws of proportion permeate the entire cosmos, and the human *microcosm* corresponds to the *macrocosm* in a model that is both mathematical and aesthetic. Boethius thus had a decisive influence on medieval thought. Here, too, Greek philosophy and views on nature are brought into line with biblical statements in their clarity: on the one hand, the Greek pre-Socrat and mathematician Phythagoras, and on the other, the wisdom of Solomon with measure and number as a divine principle that creates order. The whole numbers therefore already played a major role in antiquity, but especially in medieval allegorical texts. This will be concretised using two examples: Even for the Phythagoreans, the cosmos and all things are determined by the number *three*: Beginning, middle and end (Heinz-Mohr 1971, 309). In the Old Testament, the sacredness of three is attributed to Abraham's encounter with God, who appeared to him as a trinity: "And the Lord appeared to him [...] And he [Abraham] lifted up his eyes and saw, and behold, three men stood before him" (Gen 18:1,2). In the Christian tradition, the number three stands for the Trinity of God (Beigbeder 1998, 458; Heinz-Mohr 1971, 308 f.). The reference comes from the baptismal formula in the Gospel of Matthew: "[...] baptising them in the name of the Father and of the Son and of the Holy Spirit" (Mt 28:19). Jonah was in the

---

[16] See URL: https://digital.onb.ac.at/rep/osd/?11345682 (accessed 24/05/2024).

fish for three days (Jonah 2:1), and the resurrection of Christ is said to have taken place on the third day (1 Corinthians 15:4). Three thus becomes the number of perfection and completion in the Christian faith (Heinz-Mohr 1971, 308).

The number *four,* on the other hand, already emerged in antiquity in earthly terms, i.e. as a world number: according to the Greek natural philosopher Empedocles, the world consists of the four elements fire, water, earth and air (Aster 1968, 48-50). In the Old Testament, the Garden of Eden is fed by four streams of paradise (Gen 2:10) and there are four edges of the world (Ez 7:2). This is adopted in Christianity, as can be seen from the letter of a Carthusian monk who refers to the four parts of the world, four elements and four main winds (Eco 1993, 58). The relationship of the four to the world can be traced beyond the Middle Ages to Friedrich Schiller, who establishes the same relationship in the *Punschlied*: "Four elements/ Innig gesellt/ Bilden das Leben/ Bauen die Welt" (Friedrich Schiller Archive 2022).[17] In section 6 of this paper, further numerical allegories are discussed in context.

In connection with the significance of numbers, another aspect of Romanesque buildings must be mentioned here, which initially has no symbolic-allegorical references, but rather describes a pragmatic basis of the building huts of the High Middle Ages. Albrecht Kottmann, a graduate engineer, has analysed the measurement methods of Romanesque master builders. He points out that the basis of the measurements refers either to the equilateral triangle or the square, i.e. simple geometric figures that could be quickly drawn anywhere using a compass with a given radius (Kottmann 1971, 14-17). In such a figure, many different dimensions could then be defined, for example by triangle height or doubling to a hexagon,

---

[17] The corresponding URL can be found in the bibliography under *Friedrich Schiller Archive.*

in the case of a square by diagonal or doubling to an octagon or by further simple geometric extensions of the figure, which were then the *measured* basis of a building construction, for example to determine column height, column spacing, base diameter etc. Kottmann uses around 60 Romanesque buildings to show that either the *triangulature* or the *quadrature* was used as the basis for dimensioning. For the (further) construction of Milan Cathedral, there are documents from 1391 that deal with the question of whether to continue building "ad quadratum" or "ad triangulum" (Kottmann 1971, 14; Binding 1993, 262). Kottmann also presents calculations on the Wimpfener Pfalz, which support the thesis of the present work and are explained in section 6.

4. Medieval architecture as a carrier of meaning

The art historian and philosopher Günter Bandmann points out that superordinate meanings in the Middle Ages intervened in the development of a building in a way that we have yet to discover again today (Bandmann 1998, 10-12). From the philosophical, theological and aesthetic foundations presented above, the search for indications of meaning in important buildings is almost inevitable. These meanings thus do not refer to the statics of a building, not even to an artistic meaning, but rather they give a meaning to the...

[...] a reference to something that goes beyond the material and formal organisation of the artwork, a classification in a larger context of meaning, [...] by understanding the artwork as a parable, as a representation, as a material emanation of another. (Bandmann 1998, 11).

This can be harmonised precisely with the ideas of scholasticism explained above, that the *true is* revealed in the *beautiful*. According to Bandmann, the Platonic philosophy of an *idea* behind sensory perception is also expressed in medieval architecture:

Because of the instrumental character of the "significant" and "meaningful" forms, recognition by the viewer is not a naïve act, but originally represents a high achievement of the intellect, since the overriding concept - the idea - of the object must be present in the viewer in order to be able to determine and classify the pictorial appearance. [...] While we delectate the stoniness of the material and allow ourselves to be influenced by the naivety and freshness of the spatial composition, in the Middle Ages it was precisely the sublimation of the material, the integration of the elements into the speculative structure of meaning [...] that was exciting." (Bandmann 1998, 12 f. and 25, inverted commas by the author).

According to Bandmann, the choice of forms is therefore not related to the design or a claim to the *beauty of* the forms, but the *meaning* of the forms is at the centre. As mentioned several times by Eco, Bandmann also focuses on the "symbolic-suggestive" (Bandmann 1998, 15) aspects of the medieval building and an "urge to allegorise" (Bandmann 1998, 25). However, these approaches were also criticised: According to the art historian Günther Binding, since Bandmann the question remains unanswered "[...] to what extent architectural allegorisation determined the choice of building forms [...]", the interpretation could also have taken place retrospectively (Binding 1998, 382 f.). However, Eco had already pointed out the problem that the symbolic-allegorical way of thinking was difficult to convey to rational people. Bandmann himself also comments on such criticism: the sense for indicative - according to Eco, *symbolic-allegorical* - architectural elements was lost from the 19th century onwards. However, this change should not lead to the architecture of the past being analysed under a recent aesthetic (Bandmann 1998, 24). Bandmann also addresses the significance of symbolic numbers. For example, the number of supporting elements can provide symbolic indications, but the possibilities of architectural, numerically symbolic carriers of meaning are even more subtle:

The symbolic number can also be related to the dimensions of the building elements and take the place of the ancient proportions determined by formal laws. Three or four repeated dimensions [...] suffice to ensure the identity of meaning" (Bandmann 1998, 48).

Before analysing the building structures of the Königspfalz Wimpfen under these aspects of meaning, the necessary semiotic theoretical framework will be explained briefly.

5. An introduction to semiotics

The symbolic-allegorical interpretation of building structures that follows in section 6 must not be based on arbitrary associations in a scientific work. Building on the empirical historical findings and the contextual theological and philosophical background that emerges from the literature, the attributions of meaning are based on the theoretical framework of semiotics. A brief overview is required. This theory deals with phenomena that are generally understood as *signs*. Generally speaking, semiotics examines everything that stands for "something else" (Chandler 2007, p.2). Going back to Ferdinand de Saussure, a distinction is made in linguistic communication between the *signifier,* the sound of a spoken word, and the *signified,* which represents the idea of what is meant on a purely mental level. Signifier and signified are understood together as a *sign* (Chandler 2007 14 f.).[18] In Saussure's reception, the signifier as the sound of language was later also applied to the written word. The signs of such a signification function within the framework of a social agreement in the sense of a code and thus become carriers of meaning (Chandler 2007, 147). When we hear or read the word *tree*, we mentally imagine an entity from nature with a more or less uniform structure, e.g. the image of a trunk with branches, perhaps with green leaves or needles, but in the context of an ecological discourse perhaps the image of a defoliated or dead tree. Saussure emphasised that the relationship of the signifier to the signified is not subject to any generalising inevitability, i.e. it is arbitrary (Chandler 2007, 19). This can be illustrated simply by considering the different written or

---

[18] At about the same time as Saussure's *dyadic* theory of signs (signifier and signified, whereby the latter is understood as a purely mental concept), Charles Sanders Peirce developed a *triadic* model in which the object meant, possibly tangible, was included, thus creating a *semiotic triangle*. (Chandler 2007 29-35).

phonetic signifiers for one and the same mental image - for example the signified *tree* - in different languages. In addition to arbitrariness, a central point in Saussure's work is the indication that the *meaning of* a sign - i.e. the dyad of signifier and signified - only arises from the difference to other signs. The meaning of *red* at a traffic light results from the difference to *green*. The meaning of the word *lamb* on a farm in a certain situation results from the difference to, for example, *sheep*, but on a menu perhaps as the difference to a vegetarian dish. Saussure's structuralist text analysis also points to the horizontal structure of the *syntax* and the vertical structure of the *paradigm* (Chandler 2007, 8485): A grammatically correctly constructed sentence with subject, predicate and object forms a horizontally readable syntagmatic narrative. On a vertical level, the subject, predicate or object can be replaced by another signifier, which changes the narrative. Saussure's theory of signs was already envisaged by *structuralism and* was later transferred beyond language to artefacts or rituals, and ultimately to all cultural phenomena that can be *read* analogously to a language (Hall 2011, 36). The Saussurean characteristics of arbitrariness in the relation of the signifier to the signified are retained, as is the attribution of meaning to a sign by differentiating it from other signs. As an illustrative example of

An example is given of the vertical paradigmatic versus horizontal syntagmatic analysis and the attribution of meaning through differentiation: The significance of a Romanesque church for people of faith can be analysed by contrasting its architecture with that of a Gothic church: The small arched windows of the rather low Romanesque churches allow little light into the room, making the additional visual impressions more difficult to perceive; one can assume a certain sensory deprivation. Perhaps this achieves or at least facilitates an inner contemplation. When

worshippers entered a Gothic cathedral for the first time in the 13th or 14th century, the high Gothic windows, the light conditions created by stained glass and the subjective perception of the proportions of their own bodies in relation to the high Gothic church interior triggered completely different sensations with consecutive attributions of meaning. The faithful may associate the light with Jesus and the splendour of colour with a divine world, the stained glass paintings impose themselves as narratives. Furthermore, the people themselves in the high space feel small in relation to the divine power. The juxtaposition of the structures mentioned here as examples corresponds to a *vertical paradigmatic correlation*,
which
attributions of meaning were analysed by looking at differences: The Romanesque or Gothic building was itself considered as a signifier and through the mutual mental exchange a different signification could be worked out, which is intersubjectively comprehensible. The observation of monsters on the outside of the west portal of a church with subsequent entry into the church interior and the passage towards the altar with a representation of the cross may lead to a significant *syntagmatic narrative* for Christian believers, namely the liberation from all evil and redemption through the sacrificial death of Jesus, the aspect of difference lies in the contrast between the monsters in front of the west entrance and the symbol of the cross facing east within the church interior. The guidance of the body in and through the space can be *read* horizontally syntagmatically as a narration. Saussure's structuralism was expanded in post-structuralism by, among other things, critically incorporating historicity: The respective structures and their attributions change in a culturally contingent way, and the influential power structures were also increasingly moved to the centre (Münker and Roesler 2000 21-35). However, contingency and the above-

mentioned arbitrariness of *signs* do not mean that the attributions of meaning are arbitrary, on the contrary: they can certainly be analysed as a code from the respective historical and socio-cultural contexts and the aforementioned semiotic aspect of difference.

In anticipation of the following analysis of building structures, according to semiotic theory - in relation to a specific number, for example - its meaning cannot be read from itself alone; it only emerges as an aspect of difference in differentiation from other numbers in the corresponding spatial context, naturally also in relation to the historical-sociocultural framework. The numerical rhythm of a row of windows or arcades can be read horizontally in a syntagmatic chain, possibly creating a narrative, as will be shown.

Based on the philosophical, theological and aesthetic context of the High Middle Ages, the following chapter from 6.3 onwards will analyse the building elements of the palace under possible aspects of meaning. The historical contextualisation of the client for the construction of this palace, Frederick I, will be an additional source of interpretation.

6. The Palatinate of Wimpfen

6.1 Nomenclature and construction time

Firstly, a note on terminology must be made: It is historically correct to call the palaces[19] on the territory of Germany *royal palaces* and not *imperial palaces*, as the Frankish rulers of the medieval empire were first and foremost kings, while the imperial title represented an additional dignity that not all kings attained (Binding 1996, 25). For the Palatinate of Wimpfen, which was built under the Hohenstaufen Emperor Frederick I Barbarossa (Ahrens and Bührlen 1980, 10 f.), the incorrect term *imperial palace has been* established since the 19th century - as for other palatinates on German soil (Arens 1982, 47, Schlag 1940, title). The palace was previously thought to have been built at the end of the 12th century; Emperor Frederick I is documented to have stayed there in 1182, as did Henry VI in 1190 and 1194 and Frederick II in 1218 and 1235 (Binding 1996, 348-351). More recent excavations by Hans-Heinz Hartmann date the construction period to 11601170 (Haberhauer and Hartmann 2009, 33). This clarification will be of importance in section 6.3 for the interpretation of the palace's arcade structure and window arrangement.

6.2 The historical context of the construction period

The construction period of the imperial palace coincided with Frederick I's five campaigns in Italy against the northern Italian cities between 1154 and 1176. After the death of Pope Hadrian IV in 1159, a split in the College of Cardinals led to the election of two popes: Pope Victor IV, in contrast to the antipope Alexander III, who was in favour of the Normans who ruled

---

[19] The term *palace goes* back to the *palatium*, the seat of the Roman emperors on the *palatine* hill in Rome. However, the term *palace is* used arbitrarily as the residence of the Frankish kings. An overview of this can be found in Binding 1996, 21-26, see bibliography.

southern Italy and Sicily, felt committed to the empire and thus to Frederick I (Görich 2006, 5059). At the Council of Pavia convened by the emperor, the imperial episcopate, under pressure from the Hohenstaufen, decided to recognise Victor IV exclusively, whereupon Alexander III excommunicated the emperor. Even when Victor IV died in 1164, this papal schism was not ended by the swift elevation of Paschalis III by Barbarossa's chancellor Rainald von Dassel. It was only after the emperor's defeat at the Battle of Legano against the Lombard cities in 1176 and the resulting truce in Venice in 1177 that Alexander III lifted the excommunication and the emperor paid him the recognising *service of* the stirrup in front of St Mark's Church. The construction period of the imperial palace was thus overshadowed by the unclear occupation of the Holy See and the excommunication of Frederick I. This conflict finds an analogue in the symbolic-allegorical structures of the palace complex, as can be seen in section 6.3.

The emperor's relationship with the Christian abbess and mystic Hildegard of Bingen is linked to Frederick I's conflicts with Rome. An exchange of letters between the two is documented, beginning shortly after the emperor's coronation (Pernoud 1996, 68 f.). The correspondence reveals a mutual respect, with Hildegard calling him an "attractive personality" (Pernould 1996, 70). However, in her last work *Liber divinorum operum,* which she began writing in 1163, she critically mentions "[...] the suppression of the Apostolic See under Frederick, the Roman Emperor, [who] had not yet come to rest." (Heieck 1998, 15). (Heieck 1998, 15). Frederick I invites Hildegard by letter to visit his palace of Ingelheim. However, a subsequent meeting there is not historically confirmed (Pernould 1996, 68 f.). Irrespective of the personal encounter in question, it is nevertheless obvious that the emperor sought the advice of such a popular abbess during the time of his excommunication, as it allowed him to demonstrate his closeness to the Christian faith. This relationship with Hildegard of Bingen is also expressed in the arcade structures, as will be shown.

Before a semiotic analysis of the Wimpfen palace structures is carried out, the relationship between the author or client of a medieval building on the one hand and the building trade and stonemasons on the other must be briefly problematised: In this relationship, many questions are still unresolved, as the art historian and archaeologist Wolfgang Metternich emphasises (Metternich 2008, 129). However, Binding points out that it was not so much the artist as the client who was responsible for the meaningful attribution of meaning, which was then to be realised by the hands of the artists or craftsmen (Binding 1998, 9). Through the choice of builders, the 12th-century patron would "[...] determine the direction of his intentions [...]" (Bandmann 1998, 46). Initially, it was certainly a question of taking inspiration from the building styles of earlier rulers. The architecture of Frederick I's time characteristically utilised the forms of late Romanesque church architecture, which is evident, for example, in the round arch and column structures (Boockmann 1998, 115); Boockmann refers in this regard to the palace in Gelnhausen. However, the form to be received is never completely captured, the model is "[...] broken down into typical parts according to the meaning, and these are regrouped in the copy." (Binding 1998, 48).

The hilltop town of Wimpfen was certainly not founded by the Hohenstaufen dynasty, but the Palatinate building dates back to the Hohenstaufen period (Knoch 1983, 353). Research into the time of construction and the stay of Frederick I in 1182 allow him to be identified as the builder (see 6.1). However, there is no written record of the

Palatinate Building itself with regard to the organisation of the construction process. The following analysis is based on Binding's considerations that the builder Frederick I influenced the attribution of meaning via the building huts.

6.3 <u>Semiotic analysis of building structures of the Wimpfener Königspfalz</u>

6.3.1 Location and basic structure

Figure 1 shows the site plan of the components that are the focus of the work. They are located on the northern edge of the so-called Eulenberg in the direction of the Neckar valley. The preserved northern front of the otherwise no longer preserved palace is marked with the red arrow, to the east of which the preserved palatine chapel can be seen (the original round-arched apse is no longer preserved).

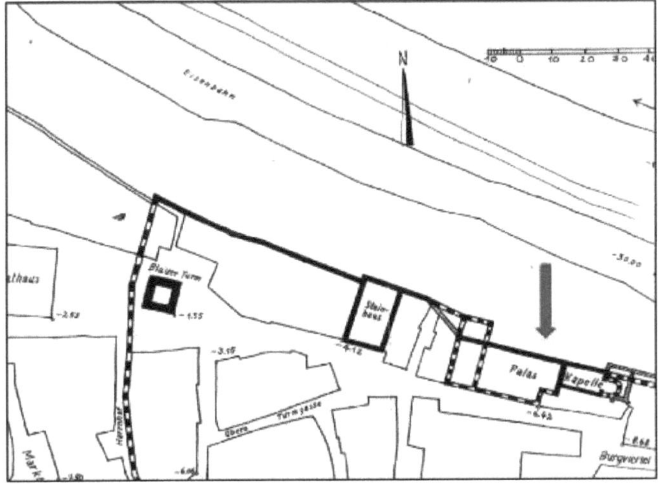

Fig. 1 Site plan, (excerpt from: Arens 1967, plate 1 in the appendix) The red arrow was added by the author.

The first reason to examine the Palatinate semiotically from a symbolic-allegorical point of view was a striking numerical rhythm of the arcade front and an equally striking column construction, which appears to require explanation in accordance with the historical religious and philosophical world view.

Fig. 2 Arcade front of the Wimpfener Pfalz, north side, photo: author

On the left (from the east), Figure 2 first shows the north front of the Palatine Chapel. This is followed to the right (westwards) by the arcaded front of the royal palace with one group of four and two groups of five arcades separated by pillars. Finally, a double window with a central column can be seen on the palace front. The pillars cannot be explained by possible partition walls in the palace building; the arcade structure delimits the large hall on the outside, in which the government office of the king or emperor was exercised with his servants, envoys and messengers (Haberhauer 2008, 69). At a time when building structures functioned as carriers of meaning and measure and number represented elements of order in the world, this 4-5-5 rhythm of the

arcade front has an interpretative prompting character. Since the row of windows of the chapel are on the same level and face the Neckar valley in a front with the arcades, the semiotic approach should begin from the east with the chapel according to a horizontal syntax and under Saussure's aspects of difference.

6.3.2 Palatinate Chapel

Here is the view of the chapel from the south front[20] :

Fig. 3 Palatine Chapel (south front), photo: author

The single-nave church - dedicated to St Nicholas - has three round-arched windows on the north and south fronts, the building widens slightly to the west and instead of a single window there are

[20] The current state results from a restoration between 1908 and 1911, as the chapel was converted into a residential building with stables in 1837. Details on this: Arens 1967, 57-72.

Also on the north and south fronts was a double window with a centre column. Behind the double window was the private gallery of the emperor and empress, to which the ruling couple had direct access from the palace. Part of the west wall of the chapel was therefore also the east wall of the palace (Arens 1967, 68), see also Fig. 1. A number three on a church building in the High Middle Ages refers to divine perfection and the Trinity, as shown above. The double window contrasts the three with a structure of two. From the Patristic period onwards, the number *two* was associated with the divine and human nature of Christ (Sachs, Badstüber and Neumann 1988, 373).[20] The syntagmatic sequence of a three with a two structure in a high medieval church can thus be read symbolically and allegorically as a sign that emically points to Christ. The fact that this aspect is not a recent interpretation can initially be plausibilised by a supra-regional art-historical finding: In the High Middle Ages, Christ was often depicted with the hand blessing, whereby, depending on the church tradition, a juxtaposition of two fingers to the other three fingers of the hand is characterised by an extension or curvature, exemplified at the time of the Normans in Monreale Cathedral (built 1174-1185):

Fig. 4, Apse of Monreale Cathedral, photo: author

A *speaking gesture* has been handed down from Roman antiquity in which the thumb, index and middle fingers of the right hand are extended and the last two fingers are curled. This is known as the *Latin gesture* (Onasch

---

[20] The divine *and* human nature of Christ was adopted as dogma at the Fourth Council of Chalcedon in 451. This decision was formative for the further theological development of Christianity (Franzen 2006, 88 f.).

1981, 132).[21] This hand position was adopted in Christianity in various modifications as a *gesture of blessing*. The *Latin gesture* is contrasted with the *Greek gesture*, which underwent a special development in Byzantium and Russia (Onasch 1981, 132): A "litteral" cross was depicted as a gesture of blessing (Onasch 1981, 220), whereby the Greek χ (chi) - referring to Christ - is symbolised by crossing the index and middle fingers; similarly, the thumb crosses the ring finger, and the little finger is curved in like the ring finger. The depiction of Monreale seems to correspond to this variant. Irrespective of these different Greek and Latin variants, the number two is symbolically contrasted with three in the gesture of blessing. As described above, two is assigned to divine and human nature, three to the Trinity (Sachs, Badstübner and Neumann 1988, 373; Onasch 1981, 220). In painting, this symbolism is still quoted centuries later in the *Latin gesture*, for example in Leonardo da Vinci's Infant Jesus in his *Madonna in the Grotto.*[22]

This Christian gesture of blessing is now also related to the Palatine Chapel in Wimpfen: the master builders of a Gothic remodelling of the choir of the Palatine Chapel in Wimpfen paid tribute to this aspect, as can be seen from a preserved keystone that can still be seen in the chapel (now the Church History Museum):

---

[21] I would like to thank Katharina Flügel, Professor Emeritus of the Saxon Academy of Science, Leipzig, for her advice and literature on the Roman gesture of oration and its different reception in the Christian gesture of blessing.

[22] See URL Louvre, Paris: https://www.pariscityvision.com/en/paris/museums/louvre-museum/madonna-of-the-rocks (accessed on 08/08/2022).

Fig. 5 Blessing hand, Palatine Chapel Wimpfen, photo: author

Although the keystone is weathered, the outstretched index and middle fingers are clearly recognisable. The 3-2 numerical rhythm can thus be found in the Palatine Chapel on the outside of the window front and inside in clearly Christian symbolism, referring to the Trinity and the divine-human nature of Jesus.

It should also be noted that, according to Kottmann, the entrance to the chapel was calculated according to the principle of triangulation, i.e. the equilateral triangle served as a measure (Kottmann 1971, 205), which can be read as an indication of the Trinity, as can be explained below from further data provided by Kottmann.

The fact that the 3-2 rhythm of the row of windows symbolises the Christian reference interpreted here can of course be criticised as speculative in itself. However, according to the semiotic theory outlined above, this meaning can be plausibilised in the further syntax of the arcade row according to the semiotic aspect of difference.

6.3.3 Palace

The palace building follows to the west as the residence and place of government of the king or emperor, i.e. the representative of secular power. As shown above, only the north front of the palace with the arcade structure still exists. Looking at the front of this imperial section of the building (see Fig. 2), the arcades are divided into 3 groups of arches by 2 pillars, as already mentioned, resulting in a 4-5-5 rhythm. The first 4-pillar structure, in contrast to the following two 5-pillar structures, cannot be attributed to any later changes. As mentioned above, the east wall of the palace also forms the west front of the chapel; both building sections were constructed as a unit. In an era in which measure and number represent a system of order, randomness cannot be assumed. This 4-5-5 rhythm of the arcades must be semiotically contrasted with the 3-2 rhythm of the chapel windows; the aspect of difference leads to the attribution of meaning: as shown above, the number four is assigned to the earthly world. In terms of numerical allegory, this is stringent in relation to the imperial palace building as a place where worldly power is exercised. The literature on Christian iconography indicates that the juxtaposition of a structure of three and a structure of four can be the deliberate basis of architectural compositions in order to refer to the symbolic juxtaposition of the divine Trinity and the earthly world (Sachs, Badstübner and Neumann 1988, 146). There is also a correlation with the data that prove the *quadrature* as the basis of the measurements on the arcades in contrast to the *triangulation* on the Palatine Chapel (Kottmann 1971, 203-205). Kottmann himself finds this fact in Wimpfen "astonishing [...]" (Kottmann 1971, 205), but the analysis of meaning presented here provides an explanation for these measurements. The synopsis of the available empirical findings in accordance with the high medieval way of thinking already points to a conscious decision by the builders in connection with the numerical allegory. The subsequent double 5 of the arcades can be interpreted from the historical context, reflecting the influence of Hildegard von Bingen: Umberto Eco emphasises the significance of the 5 in relation to Hildegard. "The mysticism of St Hildegard [...] is based on the symbolism of proportion and the mysterious magic of the five" (Eco 1993, 77). In her last work *Liber divinorum operum,* the abbess and mystic - like Vitruvius and later Leonardo da Vinci - related the proportions of man as a microcosm to the macrocosm (Heieck 1998, 91-97; 216). The *number five plays* a decisive role here (Heieck 1998, 257, Heinz-Mohr 1971, 310). Just as Hildegard divides the earth into five areas, she divides the human being vertically and horizontally into five body parts each: Head, upper body, abdomen thigh and lower leg on the one hand, twice forearm and upper arm, upper body in between on the other. According to Heinz-Mohr, "rows of five in Romanesque sculptures must therefore be regarded as symbolic and by no means as merely >>decorative<<" (Heinz-Mohr 1971, 310, emphasis added by the author). Hildegard von Bingen wrote the above-mentioned work between 1163 and 1170 (Heieck 1998, blurb).

This corresponds exactly to the construction period of the Wimpfener Pfalz, and Hildegard's close relationship to Frederick I Barbarossa has already been described above. In the *Liber divinorum operum,* the relationship between man and the world is at the centre, so as a numerical allegory, *five* is used twice for man in relation to the world number *four.* The 4-5-5 numerical rhythm on the secular palace in a horizontally consecutive sequence to the 3-2 rhythm of the chapel with its Christian connotations can be read contextually as a symbolic allegorical meaning under the semiotic theory of the aspect of difference: The numerical allegory refers to the microcosm-macrocosm relationship, which the secular ruler expressed in reference to his Christian counsellor Hildegard.

Hildegard was a politically interested woman, as her correspondence with Frederick I shows. Like other mystics of the time, she sought an individual relationship with God and thus bypassed the priestly role of mediator that was denied to her. She thus "questioned the hierarchical order (of the church)." (Borst 1983, 416, brackets by the author). This makes her an interesting person for Frederick I in the political situation of the dualism of empire and papacy, above all, of course, in connection with the papal schism and the excommunication by Alexander III. The emperor himself left no doubt that his dignity did not depend on his coronation by the pope: A quote from Frederick I serves as an example: "Since by the election of the

Princes the kingdom and the empire is ours alone from God [...]" (quoted from Bookmann1998, 97). 4[2]

Despite all the conflicts with the papacy, Frederick I's contact with Hildegard was certainly intended to leave no doubt as to his Christian roots in the faith. The Hohenstaufen's connection to Christianity is also evident in the pillar structure of the Palatinate as a symbol of importance. This can also be worked out paradigmatically via the difference aspect of semiotics, namely in the demarcation of the similar arcaded columns in comparison to the palace of Gelnhausen, which was also built under Frederick I Barbarossa around 10 years later, probably between 1170 and 1180 (Binding 1996, 263):

Fig. 6 Wimpfen arcade columns. Photo: Author

[24] According to Bookmann, the quotation comes from texts of Barbarossa's chancellery (Bookmann 1998, 96 f.) whose source is not specified in the section in question.

Fig. 7 Gelnhausen arcade columns. Photo: Author

It is noticeable that in Wimpfen two edge columns were attached directly to each pillar. This is neither static nor necessary from the point of view of superficial aesthetics, as can be seen in the comparison with Gelnhausen, where the double columns on the pillars are missing. If *beauty,* as explained in detail above, had to refer to the truth behind it and the order of the world is based on *measure and number*, then one must assume that the *number* of columns was intentional and must therefore also have an allegorical meaning. In Wimpfen, therefore, the group of four arcades does not have 3 but 5 pairs of columns, i.e. 10 columns, whereas the two groups of five have 2 x 12 columns. Columns are not only a guarantee of stability but, like Egyptian obelisks, were also erected in Greek and Roman times without any architectural function, as can be seen, for example, in Trajan's Column in Rome. Columns symbolise the connection between heaven and earth, between God and man (Heinz-Mohr 1971, 250 f.). In the Christian context and in relation to the Holy Scriptures, the numbers *ten* and *twice twelve* have very clear connotations, both times they stand in the context of a revelation of God to man: Moses' encounter with God on Mount Sinai led to the giving of the Ten Commandments. In the Old Testament, God's covenant with Abraham led to the 12 tribes of Israel, which, according to Christian interpretation, refer to the 12 apostles in the New Testament. According to Heinz-Mohr, the number twelve, which appears in other contexts in the Bible, is "one of the most important numbers, literally a leitmotif of the Bible" (Heinz-Mohr 1971, 312). The reference to God and the Christian faith would therefore have been demonstrated by the builder in all political disputes with the papacy on the building as a sign of significance.

Towards the west, after the row of arcades, there is another double window with a central pillar (see Fig. 2, but the window was visually much more elaborately designed than the double window of the Palatine Chapel and appears larger, although the actual window opening is the same height as the adjacent arcades. Arens also writes that the architectural design of this window clearly stands out from its surroundings. (Arens 1967, 48). Behind this window is a vaulted room, and from the comparison with the palace in Gelnhausen and the palace in Eger, both of which have a similar double window, Arens concludes that this room may have corresponded to a private chapel (Arens 1976, 49); the room could also have been used to store relics or imperial jewels (Arens and Bührlen 1979, 29). However, as the purpose of this room is speculative, we will not attempt to interpret the number of two here.

6.3.4 Aspect of column design

Fig. 8 Knotted column at the arcades. Photo: Author

It is worth mentioning here a small secondary finding with regard to symbolism and allegory, which can be seen in the columns of the palace: The column shafts are very differently designed, particularly striking are two so-called *knotted columns*, which are regarded in art history as *apotropaion*, i.e. as a sign averting disaster (Sachs, Badstübner and Neumann 1988, 44).

The basis for this may have been the so-called beast pillars.[23] Monsters interpreted as demonic, such as serpents or dragons, were often depicted intertwined on these beast pillars, with one animal's dragon grasping

---

[23] An illustrative example of this can be found in Souillac, URL:
https://upload.wikimedia.org/wikipedia/commons/1/13/Souillac%2C_Abbaye_Sainte-Marie-PM_32034.jpg (accessed 19/08/2022).

another animal's body, which was perceived as a *taming. The* abstraction of this entanglement - the knot - plausibilises the apotropaic interpretation, that antidivine powers were to be tamed (Sachs, Badstübner and Neumann 1988, 214). Knotted columns often appear - as in Wimpfen - on the north sides as the dark sides facing away from the sun; the north side is often also associated with apocalyptic events. (Beigbeder 1998, 301). A knotted column, which is interpreted in the same way, can be found on the north portal of the church of St Thomae in Merseburg (Sachs, Badstübner and Neumann 1988, 214), whose construction period coincides with the Wimpfen Palatinate.

6.3.5 The geographical *orientation* of the chapel and palace

The Palatine Chapel was already documented in 1293 as *St Nicholas' Chapel.* The historical connection lies in the union of the Hohenstaufen dynasty with the Normans who ruled southern Italy through the marriage of Henry VI, son of Barbarossa, to the Norman princess Constance in 1186 (Stürner2009, 15). The remains of St Nicholas of Myra are venerated in the Basilica of St Nicholas in Bari; in Norman and Hohenstaufen times, he was honoured as "[...] the most important saint in the newly acquired Lower Italy [...]" (Arens 1967, 17).

Christian churches have been *orientated to the* east, i.e. towards the sunrise, since the 5th century at the latest (Beigbeder 1998, 298 f.). On the one hand, this was justified by the fact that Jerusalem was located in the east.

One liturgical argument comes from the Gospel of John, in which Jesus is described as the "light of the world" (John 8:12). The sun is therefore also regarded as a symbol of Christ. The eastward orientation means that the first rays of the rising sun can fall into the church interior through a window in the apse. According to Christian tradition, the return of Christ was also expected from the east: "ex oriente lux" (Beigbeder 1998, 299). However, many medieval churches are not orientated to the exact geographical east. The eastern orientation often corresponds to the direction in which the sun rises on the horizon on the church day of the patron saint of this church (Beigbeder 1998, 299). The fact that the orientation of medieval churches according to their patron saint's day was quite common is also proven by a work that also deals with the possibilities of medieval measuring methods and the mathematical foundations (Eckstein, Büll and Hörnig 1995, passim). In this context, a significant aspect may be added to the eastward orientation of the Palatine Chapel of Wimpfen - and thus also of the palace building, which was built in line with the chapel: Otto Scriba describes the orientation of the chapel as deviating by 10° to the south (see also site plan, Fig. 1) This axial deviation coincides with the sunrise on 6 December, St Nicholas' Day, which could be expected according to the patrocinium (Scriba 1924, 58). According to his publication, Scriba had the measurements carried out by mathematician friends, but there is no calculation data. Martin Kieß, who for years measured medieval churches with a working group from the Ludwig Uhland Gymnasium in Kirchheim/Teck, also carried out calculations for the Palatinate Chapel in Wimpfen in 2009/2010. Of course, the replacement of the

Julian by the Gregorian calendar. In 150 churches surveyed by this working group to date, it was found that in only ten per cent did the building line refer to the day of the respective church patron saint; most of the orientations referred to other important saints (Kostner 2010). According to his previous findings, Kieß also assumes that it was not the easting *to be calculated that* was decisive for the building lines, but the sunrise according to the topographical location: surrounding hills etc. make the sunrise of a particular church location appear correspondingly delayed. On the one hand, this relativises the measurement techniques presented by Eckstein et al. 1995, but on the other hand it is more comprehensible: the measurement and calculation possibilities for determining the geographical east or the geographical sunrise of a particular church day were certainly not available at every place at every time, and the topographical approach also makes the sunrise on the respective day sensually comprehensible for people in the respective church. The measurements taken by the working group led by Kieß showed that the Palatinate Chapel in Wimpfen was oriented to 22 February. This day is known in the church calendar of the Roman Catholic Church as the *Cathedra of Peter* and celebrates the appointment of the Apostle Peter to the teaching office of Roman bishop. The popes are regarded as his successors in an uninterrupted apostolic succession (Franzen 2006, 109 f.). The building period during Barbarossa's excommunication, in which the occupation of the Holy See was of decisive importance for Frederick I, thus finds a correlate in the building line. The fact that the building line is only visible to the observer in exceptional situations (from the western defence tower - today's Blue Tower - it would have been visible on the

22 February), plays no role according to medieval philosophy and aesthetics. As explained in detail above, there is a truth behind the superficial perception of the senses. The "dialectic between idea and reality appears as a dialectic between the thing and its essence" (Eco 1993, 129). At the cathedral in Reims, for example, there are highly elaborated royal figures high up on the buttresses, the fineness of which cannot even be remotely perceived by onlookers; the finished work was only visible "to the eye of God" (Pinder 1992, 53)[24] . In this - from our perspective - pre-

____

[24] I would like to thank Tobias Frese from the Institute for European Art History at the

scientific world of faith, artefacts worked in their pure existence, as can also be seen in the attributed demon-defending power of the knotted columns. If the orientation of the Palatine Chapel and thus also of the palace to the *celebration of Peter's chair can* be attributed to a conscious decision by Frederick I, the underlying idea remains speculative for us. However, the coincidence of the orientation seems at least questionable in relation to the historical context.

University of Heidelberg for this comment on the phenomenon of "illegibility" and for further suggestions on the aesthetics of reception in art history.

7 Summary and critical evaluation

The conspicuous numerical rhythm of the palace chapel windows and the row of arcades of the Wimpfen royal palace, the equally conspicuous order of columns and the ostentation of these building structures were worked out as carriers of meaning in the present work. The historical and socio-cultural context served as a starting point, with philosophical, theological and aesthetic thought patterns at the time of construction playing a central role in the interpretation. The literature on medieval buildings as carriers of meaning was used to legitimise the project. The theoretical framework of semiotics was used as an analytical method. Saussure's semiotics emphasises that the meaning of a sign does not rest in itself, but can only be inferred from the difference and juxtaposition to other contextual signs on a horizontal level. Both architecture and the statues, reliefs and ornaments of the High Middle Ages can then be *read* semiotically as a narrative:

It is important to learn to read these pictorial works. Reading these images is achieved by first learning and understanding the letters, then the words and finally whole sentences, as in literature [...] If one understands how to read these images, the Middle Ages give the viewer completely new insights into their lifeworlds from a different perspective (Metternich 2008, 9).

A further argument in support of the thesis may be added to the historical socio-cultural framework: In 1999, the German Research Foundation (DFG) approved a research training group at the University of Münster on the subject of "Social Symbolism in the Middle Ages" (Westfälische

Wilhelms-Universität Münster 2000).[25] In awareness of the recent problems in communication processes with other cultural social systems, the Middle Ages are used as a paradigm:

The plurality of cultures made the problem of understanding and translation virulent early on and gave rise to an awareness [sic!] in dealing with formalised communication and a sensitivity for the legibility of signs that was not necessary in simpler, homogeneous societies. At the same time, the system of allegorical interpretation of the world, man, history and society developed from Jewish patristic biblical exegesis impregnated all areas of life with transcendental significance, which lent the symbols an ontological dignity (Westfälische Wilhelms- Universität Münster 2000).

From this research project alone, which was extended by the DFG until 2005, it is possible to deduce the importance that must be attributed to the reading of nationally recognised symbols even today. And people today are virtually inundated with *symbols* in the form of *icons*: from traffic signs to toilet instructions and warnings about the dangers of toxic substances to *icons* in the digital world.

In criticism of my thesis, it must be clearly stated that there is no evidence for this reading of the analysed building structures from the historical context. Günther Binding has already been quoted with his criticism that allegory could also be read into the buildings as meaning retrospectively (Binding 1998, 382 f.). However, one argument against this criticism has already been mentioned by both Umberto Eco and Günther Bandmann: medieval aesthetics are initially alien to us, but we cannot and must not analyse the architecture of the past under a recent aesthetic (Bandmann 1998, 24).

In my opinion, the interpretation presented in this work is strongly

---

plausible on the basis of the aforementioned conspicuous features and the corresponding context. According to Eco, the most characteristic aspect of medieval aesthetic sensitisation was the symbolic-allegorical view of the world, a striving for interpretation (Eco 1993, 79-83). According to Bandmann, there is an "urge to allegorise", to understand buildings in terms of their meaning (Bandmann 1998, 25). But even assuming that all the empirically tangible numerical structures and forms described in this work are a product of chance, we are closer to the thinking and life of medieval man in the context-orientated *attempt at interpretation* than with purely documentable historical facts.

Literature

Arens Fritz (1967): Die Königspfalz Wimpfen, Deutscher Verlag für Kunstwissenschaft GmbH, Berlin.

Arens Fritz and Bührlen Reinhold (1980): Wimpfen: Geschichte und Kunstdenkmäler, self-published by the Alt-Wimpfen Association, Bad Wimpfen.

Arens Fritz (1982): Wimpfen as the new centre of Hohenstaufen power on the lower Neckar, in: Regia Wimpina: Beiträge zur Wimpfener Geschichte, Verein "Alt Wimpfen" (ed.), pp. 39-56, self-published by the Alt-Wimpfen Association, Bad Wimpfen.

Aster Ernst von (1968): History of Philosophy, Kröner, Stuttgart.

Bandmann Günter (1998): Mittelalterliche Architektur als Bedeutungsträger, Gebrüder Mann, Berlin.

Beigbeder Oliver (1998): Lexicon of Symbols: Schlüsselbegriffe zur Bildwelt der romanischen Kunst, Echter, Würzburg.

Binding Günther (1993): Baubetrieb im Mittelalter, Wissenschaftliche Buchgesellschaft, Darmstadt.

Binding Günther (1996): Deutsche Königspfalzen von Karl dem Grossen bis Friedrich II. (765-1240), Wissenschaftliche Buchgesellschaft, Darmstadt.

Binding Günther (1998): Der früh- und hochmittelalterliche Bauherr als sapiens architectus, Wissenschaftliche Buchgesellschaft, Darmstadt.

Bookmann Hartmut (1998): Stauferzeit und spätes Mittelalter: Deutschland 1125-1517, Siedler, Berlin.

Borst Otto (1983): Everyday Life in the Middle Ages, Insel, Frankfurt am Main.

Cancik Hubert and Mohr Hubert (1988): Religionsästhetik, in: Hubert Cancik et al. (eds.): Handbuch religionswissenschaftlicher Grundbegriffe,

Kohlhammer, Stuttgart, pp. 121-156.

Chandler Daniel (2007). *Semiotics: The Basics*. Routledge. London, New York.

The Bible (1999): After the Translation of Martin Luther, With Apocrypha, German Bible Society (ed.), Stuttgart.

Diemer Alwin (1970): Ontologie, in: Alwin Diemer and Ivo Frenzel: Philosophie, Fischer Bücherei, Frankfurt am Main/Hamburg, pp. 209-240.

Dollmann Jürgen (2021): Spiritual discourses and performances in German Ayurveda centres: Transdisciplinary, integrative analysis of a "holistic" therapy procedure, online resource, Heidelberg. URL: https://archiv.ub.uni-heidelberg.de/volltextserver/30648/ (accessed 23/08/2022).

Eckstein Rudolf, Büll Franziskus, Hörnig Dieter (1995): Die Ostung mittelalterlicher Klosterkirchen des Benediktiner- und Zisterzienserordens, in: Studien und Mitteilungen zur Geschichte des Benediktinerordens und seiner Zweige, vol. 106, issue 1, 1995, EOS, ST Ottilien, pp. 7-78.

Eco Umberto (1993): Kunst und Schönheit im Mittelalter, 6th edition 2004, Deutscher Taschenbuch Verlag GmbH & Co, Munich.

Elberfelder Studienbibel mit Sprachschlüssel (2005), text no. 21, 1st edition 2005, Brockhaus, Wuppertal.

Franzen August (2006): Kleine Kirchengeschichte, Herder, Freiburg im Breisgau.

Friedrich Schiller Archive (2022): Punschlied, URL: https://www.friedrich-schiller-archiv.de/gedichte-schillers/kurze-gedichte/punschlied/ (accessed on 06/08/2022).

Gadamer Hans-Georg (1965): Philosophisches Lesebuch, Fischer, Frankfurt am Main and Frankfurt.

Görich Knut (2006): Die Staufer, Herrscher und Reich, Munich.

Haberhauer Günther (2008): Die Staufer und ihre Pfalz zu Wimpfen, Verein "Alt Wimpfen" e.V. (ed.), Bad Wimpfen.

Haberhauer Günther, and Hartmann Hans-Heinz (2009): Neue archäologische Erkenntnisse zur Baugeschichte der Königspfalz Wimpfen, in: Kraichgau: Beiträge zur Landschafts- und Heimatforschung Folge 21, Heimatverein Kraichgau (ed.), pp. 17-33, Eigenverlag des Heimatvereins Kraichgau, Eppingen.

Hall Stuart (2011): The Work of Representation. In Stuart Hall (ed.): Representation: Cultural Representations and Signifying Practices, p.1374), SAGE, London et al.

Heieck Mechthild (1998, ed./trans.): Hildegard of Bingen: Das Buch vom Wirken Gottes, Liber divinorum operum, Pattloch, Augsburg.

Heinz-Mohr Gerd (1971): Lexikon der Symbole: Bilder und Zeichen der christlichen Kunst, Eugen Dietrichs, Cologne.

Knoch Peter (1983): Die Errichtung der Pfalz Wimpfen - Überlegungen zum Stand der Forschung, in: Landesdenkmalamt Baden-Württemberg (ed.): Forschungen und Berichte der Archäologie des Mittelalters in Baden-Württemberg, vol. 8, Stuttgart, 343-367).

Kostner Claudia (2010): Die Heiligen und der Sonnenaufgang, in: Heilbronner Stimme, Jahrgang 65 Nr. 186, 14 August 2010, Heilbronner Stimme GmbH & Co KG, Heilbronn, p. 38.

Kottmann Albrecht (1971): Das Geheimnis romanischer Bauten, Hoffmann, Stuttgart.

Metternich Wolfgang (2008): Bildhauerkunst des Mittelalters: Botschaften in Stein, Wissenschaftliche Buchgesellschaft, Darmstadt.

Münker Stefan and Roesler Alexander (2000): Poststructuralism, Metzler, Weimar.

Onasch Konrad (1981): Liturgie und Kunst der Ostkirche in Stichworten:

unter Berücksichtigung der Alten Kirche, Koehler & Amelang, Leipzig.

Papineau David (2006): Philosophie: Eine illustrierte Reise durch das Denken, Nikolaus de Palézieux (transl.), Wissenschaftliche Buchgesellschaft, Darmstadt.

Pernoud Régine (1996): Hildegard of Bingen: Her World - Her Work - Her Vision, Herder, Freiburg im Breisgau.

Pinder Wilhelm (1992): Die Anerkennung des Betrachters, in: Wolfgang Kemp (ed.): Der Betrachter im Bild. Kunstwissenschaft und Rezeptionsästhetik, Reimer, Berlin, pp. 51-59.

Poeschel Sabine (2007). Handbuch der Ikonographie: Sakrale und profane Themen der bildenden Kunst, Wissenschaftliche Buchgesellschaft, Darmstadt.

Sachs Hannelore, Badstüber Ernst, Neumann Helga (1988): Christliche Ikonographie in Stichworten, Koehler & Amelang, Leipzig.

Schlag Gottfried (1940): Die deutschen Kaiserpfalzen, Klostermann, Frankfurt am Main.

Scriba Otto (1924): Wimpfen am Neckar: Pictures from history and art, Salzer, Heilbronn.

Stürner Wolfgang (2009): Friedrich II. 1194-1250, Wissenschaftliche Buchgesellschaft, Darmstadt.

Westfälische Wilhelms-Universität Münster (2000): Annual Report 1999, URL: https://www.uni-muenster.de/Rektorat/jb99/jb9956.htm (accessed on 23/08/2022).

MIX
Papier aus verantwortungsvollen Quellen
Paper from responsible sources
FSC® C105338
FSC
www.fsc.org